KT-167-584

BEAUTY REIMAGINED

BEAUTY REIMAGINED

Life lessons on loving yourself inside and out

PENGUIN LIFE
AN IMPRINT OF
PENGUIN BOOKS

PENGUIN LIFE

UK | USA | Canada | Ireland | Australia
India | New Zealand | South Africa

Penguin Life is part of the Penguin Random House group of companies whose addresses can be found at global.penguinrandomhouse.com.

First published 2019
001

Set in 10.25/18pt Futura Std
Typeset by Jouve (UK), Milton Keynes
Printed and bound in Great Britain by Clays Ltd, Elcograf S.p.A.

A CIP catalogue record for this book is available from the British Library

ISBN: 978–0–241–38495–4

www.greenpenguin.co.uk

Penguin Random House is committed to a sustainable future for our business, our readers and our planet. This book is made from Forest Stewardship Council® certified paper.

CONTENTS

PREFACE

Can you be a feminist and love beauty?
Of *course* you can – on that, *Stylist* has been unequivocal since the brand's launch back in 2009. But that doesn't mean the approach has to be traditional. We've always strived to make our beauty content refreshing, emotional, provocative and inclusive. Our award-winning beauty team continually challenge 'traditional' beauty ideals, scrutinize the science behind every product and consider how history, popular culture and technology shape the way we feel about how we look.

It's this thoughtful approach that has won us plaudits from the industry and made readers feel 'seen' on a topic that can feel as important as politics or pay gaps. Our relationship with our appearance is a conversation we want to keep on having as we move into an era of beauty positivity and inclusiveness. Which is why we're following the hugely successful *Life Lessons from Remarkable Women*, our 2018 anthology of essays from inspirational women, by tackling the fascinating arena of beauty.

Beauty Reimagined invites eleven women, including journalist and author Caitlin Moran, actress and advocate Rose McGowan, poet Yrsa Daley-Ward, academic Mary Beard and MP Stella Creasy, to discuss what beauty means to them. Together their unique and honest essays provide a sometimes heart-warming, often challenging but always insightful look at how beauty is about so much more than a lick of mascara. Rather, it's a powerful tool of self-expression and acceptance.

For all the women who forget to be as kind to
themselves as they are to their friends

MARY BEARD

on the joyful authenticity of grey hair

One of the most original and best-known classicists working today, Mary Beard is Professor of Classics at Newnham College, Cambridge, and the classics editor of the *Times Literary Supplement*. She is a fellow of the British Academy and a member of the American Academy of Arts and Sciences. Her books include the Wolfson Prize-winning *Pompeii: The Life of a Roman Town* (2008) and the bestselling *SPQR: A History of Ancient Rome* (2015), and her popular *TLS* blog has been collected in the books *It's a Don's Life* and *All in a Don's Day*. Her latest book is *Women & Power: A Manifesto* (2017).

When a woman doesn't buy into the world of make-up, hair dye and all of the other tools that we are supposed to use to maintain a certain illusion, you sometimes hear people remark: 'So-and-so isn't at all concerned with her appearance.' But that is always wrong. There isn't a person in the world, apart from some people at the most extraordinarily challenged end of the spectrum, who is not bothered about how they look.

It all comes down to your sense of identity. The way that you look is very tied up with how you feel about being yourself. It's about feeling that your image matches up with the person that you feel you are, and everybody thinks about that. I would really like to undermine this sense that there are people who don't care about how they look. Of course they care. They just care in a different way.

Over the years, I know many people have assumed that I don't care about my appearance because I don't wear make-up. I actually used to wear make-up and, to be honest, I can't even remember why I stopped. For years now, it just hasn't seemed to matter to me enough to bother with it. The way that I look without make-up is the way that I am; it feels real to me.

It's the same with my long, grey hair, which I do feel has become my trademark. Contrary to what some of the nastier journalists might say, I do have it trimmed, and I do brush it. It may not look like it, but I do. My hair is a bit fly-away, but that's just how it is, and it feels like me now. That's how I am. I believe that this sense of identity is particularly important as you get older. For your presentation to be successful, it must represent a good accommodation between how you feel and how you look.

I have always held this point of view, and so I have never coloured my hair, even when I first started going grey. I was probably about twenty-two when I got my first grey hair. Of course, my initial reaction was panic: 'Oh my god, a grey hair, I'm getting old!' But it was never that I feared the grey hairs themselves – it was the intimations of mortality that they represent, particularly when you're in your twenties. At first, I pulled them out. I think everyone does that but, before too long, unless you want to be bald, you have to give up that struggle. You get single grey strands for ages, they take a long time to really come through, so it is years before your hair seems grey to the outside world. I look back at pictures of me in my thirties, when my children were young, and I did go through a funny stage with bits of grey. I didn't go properly grey until I was in my fifties.

Beauty and fashion are not the enemy. You just have to work with them so that you can enjoy them without being controlled by them. I don't want to feel controlled, by anything. That's not to say that

BEAUTY AND FASHION ARE NOT THE ENEMY. YOU JUST HAVE TO WORK WITH THEM SO THAT YOU CAN ENJOY THEM WITHOUT BEING CONTROLLED BY THEM. I DON'T WANT TO FEEL CONTROLLED, BY ANYTHING.

I don't enjoy experimenting. Last year, before hosting *Front Row Late* on BBC2, I put a pink streak in my hair, which I loved! One thing that's great about having grey hair is that you can colour bits of it so easily, which means you can have fun with it. And fun is exactly what it is and should be.

Other than that, I've always been very opposed to colouring my hair completely. It is partly practical, because you have to be constantly vigilant about your roots showing and always going to the salon. To me, that's not fun. That's pressure.

I don't want to pretend not to be grey, because that's what I am. It's important to add that I don't mind other people dyeing their hair, if that's what they enjoy doing. But, with my own appearance, I would prefer to find an empowering way of feeling that I look like me, and I enjoy exploiting the natural and inevitable process of ageing. When I walk down the street, that's what I want to feel like: me. I don't want to feel that I'm pretending to be someone fifteen years younger. Sixty-three-year-old women look like me, and I'm fine with that. It's as it should be.

People often think that men are off the hook when it comes to grey hair, because we describe older men in positive ways: 'a silver fox'; authoritative and powerful. But, in fact, men are caught by this, too. I discovered just how many men colour their hair when I made a radio programme about grey hair, *Glad to be Grey*, for BBC Radio 4. It's interesting because women talk about it quite a lot, particularly in conversation with other women. You wouldn't really find a woman who pretended that she didn't colour her hair. But getting men to talk about it . . . my goodness me, they haven't come out yet! The truth is that a lot of men do use a lot of help to stay sleekly black-haired, yet our producer was calling around person after person and they wouldn't talk about it. We rang up a rather posh male salon and they said they couldn't suggest anyone because absolute discretion was required. In the end, I had to persuade one of my colleagues, who I knew dyed his hair, to come and talk on the programme because we couldn't get anyone else to do it.

So, on one hand, you've got the craggy silver fox, but then you've also got guys who are colouring their hair but so very secretively. It was quite a shock to me. I always thought men either went grey and got authoritative, or they just didn't go grey at all. It was an eye-opener. Women are ahead here, at least in openness.

Of all aspects of beauty and appearance, hair is one of the most personal to me. The other is shoes, because they are wonderfully equalizing. My daughter is over thirty years younger than I am, and she often pinches my shoes. But, yes, I do only wear flats. I have a certain feminist suspicion about high heels because they are something of a ball and chain. They slow you down.

My favourite shoes are a pair that I got from Manolo Blahnik, who's a friend of mine. They're red, flat leather shoes with thick ribbons for laces. We met when he was doing a book about shoes and wanted to ask me about their history, so we went out for lunch and talked about ancient shoes and we've been friends ever since.

The history of beauty is fascinating because, when we look at women's relationship with it, in many ways it hasn't actually changed that much. There are fundamental similarities throughout the centuries, the main one being a very long history of men trying to make women look how they want them to look. Even though I can talk gaily about wanting to feel comfortable in my appearance, this still happens, even to me. That idea is a constant, and it's very deeply ingrained.

In some ways, I'm lucky because I'm pretty tough. When somebody like A. A. Gill says that I'm 'too ugly for television' and 'should be kept away from cameras altogether', I can sort of shrug it off. But over the course of history, you can see that men have tried to police women by policing their appearance, in all kinds of different ways. They have encouraged women to wear things that are very inconvenient or very imprisoning. Corsets, crinolines – and you might add high heels to that list, despite many women saying they love their heels. And the most insidious thing is that it has also happened around the idea of make-up. Women have been encouraged to change their appearance; to change their faces. To literally paint their faces. And then they have been criticized by male culture for being, somehow, a pretence. Make-up is the thing that both makes you what men want you to look like and, at the same time, allows them to criticize you for being 'not real', or inauthentic. In mainstream culture – and this goes back for centuries; there are Roman examples of this – women get caught between the

MAKE-UP IS THE THING THAT BOTH MAKES YOU WHAT MEN WANT YOU TO LOOK LIKE AND, AT THE SAME TIME, ALLOWS THEM TO CRITICIZE YOU FOR BEING 'NOT REAL', OR INAUTHENTIC.

idea of authenticity, where men say, 'You look old,' and, on the other hand, the idea of the artifice of women's appearance and the sense that they are simply a fraud. Every which way, women get caught out. They can't win.

Those kinds of issues haven't gone away, and it's difficult to see a time when they will go away. I'm quite resilient, and quite comfortable with myself and how I look. But, at the same time, I'm very aware – how could I not be? – of the endemic sexism in our culture.

Men telling women how they should or should not look has many, many varieties, right down to the burka. In that example, a woman is told, if you expose yourself, you're exposing your flesh, which is wanton and shameful. If you cover it up, you're not living the 'Western way'. All of this is very, very loaded.

How does anyone become resilient in the face of that? I wish I knew, actually. If I had a quick answer to that, I would share it with every single woman. But, unfortunately, I think it comes down to time and learning to be comfortable with yourself and feeling that you're doing what you want to in life. Certainly, that's how it was for me.

The way I think about it is this: for a very long time when I was a new lecturer, I was really very anxious. I felt uncomfortable about lecturing, as though I didn't quite know how to do it. Even contributing in seminars, I could never quite get my point in. I didn't own my own voice. Then one day, when I was in my late thirties, after I'd given a lecture, I had this odd feeling that 'That was me speaking.' It wasn't me pretending to be an academic or pretending to be one of my colleagues – male, of course. I was just, in that lecture, speaking for me. And that was a huge turning point. I don't know how it happened. I wish I did because I would tell the world. But somehow, I suddenly seemed to inhabit my own voice. And I feel a bit like that with appearance, and hair and make-up and shoes. I feel that I'm inhabiting me, which allows me to play with it and have fun, and that certainly hasn't always been the case. That confidence and resilience comes with experience. I don't think anyone is born with it.

I was an absolutely predictable teenager and twenty-something: always thinking that I was too fat, too thin, too whatever. And I believe that every woman goes through that. I can't think of anybody who has been confident with their body all throughout their entire life. In part, that is one of the advantages of getting older, if you manage to play it that way. I know that some people find the disappearance of that youthful body difficult. For me, it's been a liberation, because you're out of that immediate spotlight. I think it's something that everybody has to work out individually, and it's very hard to give somebody else advice about it.

If I had a master plan here, I would share that, too, but I think it's actually a complicated journey to feeling respected or taken seriously. Underneath all of the external elements of femininity, what women want – perhaps more than anything else, because they often don't get it – is to be taken seriously. And when you feel that you're taken seriously, it gives you much more self-ownership of how you look and how you sound and how you feel about yourself.

Now, in my sixties, I'm very happy with my appearance. Yes, there are people who think that I don't care about how I look, but nothing could be further from the truth. I'm simply putting myself into what I consider my comfort zone, not anybody else's.

So if people say, 'So-and-so isn't concerned about how she looks,' what I think they actually mean is 'She doesn't want to look how I want to look,' or 'She doesn't want to look how I want her to look.' There isn't a sentient person on the planet who doesn't have some sense of how they want to put themselves forward, or how they want to look in the world, and there are lots of different versions of that.

To be honest, I don't give a shit what other people decide to do with their own appearance. What I do mind about is when they're trapped by that. I can't bear to hear a woman saying, 'I couldn't go outside without having put my mascara on.' Then it has stopped being about how you are going to be in the world, and it's veering towards being some kind of a disguise that you put on, which gets between you and the world. You have to find a way of being and looking and presenting yourself that feels right for you.

And that's why these issues are so important and must be discussed. How people feel about how they look is absolutely crucial. It would be ridiculous to dismiss – as many 'serious' thinkers do – the concept of appearance, as if it were utterly superficial and meaningless. Instead, it's clear to me that it is central in the world and always has been.

CHARLI HOWARD

on the journey from self-loathing to self-acceptance

Charli Howard is a curve model, author and body-positivity campaigner. She splits her time between London and New York, where she has carved out a career in curve modelling, and she has fronted campaigns for leading brands including Redken, Good American, Agent Provocateur and Pat McGrath Labs. Charli's mission is to drive awareness around diversity in the fashion industry and to help people with mental health difficulties by speaking out about her own challenges.

The day I reached my desired weight of 105lb, I felt . . . *odd.*

I stepped back and forth on the scale like a madwoman, back and forth and back and forth, just to make sure the ticker wasn't deceiving me. I couldn't believe it. *Seven and a half stone!* All the years I'd spent dieting had finally paid off. At the time, I was a twenty-three-year-old model and had been starving myself on and off for nearly a decade.

I'd dreamed of this moment for years, fantasizing about walking out of the house and people doing double-takes at my ballerina-like frame; clothes looking ten times better on me than on the average person (because hey, why wouldn't they? I was finally *gorgeous*); men falling over themselves while marvelling at my beauty.

But 'odd' was the operative word here, and that initial two-minute sense of euphoria quickly subsided. As I clocked my reflection in the mirror, I didn't see perfection. I may have reached my goal weight, but I still wasn't beautiful, despite what I'd been led to believe my entire life.

Life had taught me that, in order to be happy or valued, I had to be *thin*. It had started subconsciously as a child, when friends would complain about their weight and how they didn't have a 'playground boyfriend' because of how fat they were. It started with a friend's mum constantly complaining about her weight and saying how 'jealous' she was of my mum's figure. It started with *my* mum, jogging for miles every day and eating salads in order to stay slim. It started with cartoon characters in kids' TV shows being made fun of for being fat – or, worst of all, people in your class being made fun of for being fat. I quickly learned that 'fat' was the enemy.

As a teenager, the stigma towards being fat became worse. No one wanted to be 'the fat girl'. Being fat meant you were friendless. It meant no boys fancied you. It meant you were lazy; that you were ugly and worthless. No matter what I read or who I spoke to, if you wanted to be successful, then the message was clear: DO NOT GAIN WEIGHT. The last thing I wanted to be was curvy . . . and yet I was.

I looked nothing like my mum or my sister, who were perfectly lithe. I was fourteen and a size fourteen, with boobs and hips, and as someone who reached puberty well ahead of her peers, I felt like a total freak. I envied the sporty girls who seemed to make being thin look effortless. I was jealous of girls who could have one portion of food at lunchtime without feeling hungry afterwards. Why couldn't I look the same as them?

I'd collected fashion magazines for years, but the ones in the early noughties had soon become bibles of 'thinspiration' for me – page after page of tall, frail-looking waifs wearing the most fabulous of clothes in the most exotic of locations. As a chubby teenager with acne and low self-esteem, these models were who I desperately aspired to become. And so, with the limited knowledge I possessed, I felt I'd finally cracked the code: that being thin was my gateway to happiness.

Most people remember their youth as a time of getting drunk in fields, but all I remember is diets. In a feeble attempt to feel happy for just a moment, food – or, rather, the lack of it – soon became a solace. I tried them all – low-carb diets, the 'one apple a day' diet, the baby-food diet, the 500-calories diet. And, naturally, the weight came off. Being thin got me noticed for something. Being told I looked great was an incentive to carry on and, soon, I had to learn how to maintain the low weight. Bulimia became the answer – throwing up plates of delicious food in secret in an attempt to show the world I was perfect. And it worked: I loved it when people noticed my body getting smaller or showed concern for me.

I didn't stay a teenager for ever, of course, and by the time I reached adulthood the damage had already been done. My anorexia and bulimia – though I wouldn't acknowledge the fact it *was* that – was now deeply ingrained in my everyday life, every calorie and ounce of fat accounted for. It didn't help that these backward societal beliefs of how women should look had also infiltrated the minds of women around me. The way my friends and I looked (or didn't look) became the pinnacle of conversations, our worth as human beings based solely on what dress size we wore. Perhaps he dumped me because he saw my cellulite? Perhaps if I was thinner, I'd get the job I want?

I was ashamed of my curves, and I wasn't the only one. It became clear via our conversations that we women needed to rid ourselves of the things that made us feminine as though our lives depended on it. No matter what magazine you picked up or what TV show you watched, the

PERHAPS HE DUMPED ME BECAUSE HE SAW MY CELLULITE? PERHAPS IF I WAS THINNER, I'D GET THE JOB I WANT?

overriding consensus was the same as when I'd been a child: *Being thin will make you happy.*

By the time I'd reached 105lb, my life-long ambition of becoming a model had finally come true. But rather than making me feel content in who I was, my new weight gave me an excuse to hate myself further. I detested myself. I didn't question why, despite achieving my dream, my life didn't miraculously improve. I just figured I'd have to plough my energy into becoming even thinner, until I was finally happy.

Beauty, as I soon discovered, *was* measurable, in the guise of a 34–24–34-inch ratio. Either you reached those measurements and got the jobs, or you didn't – it was as simple as that. But no matter how much I dieted, I could never reach the standards the fashion industry said were 'ideal', and I often wondered if I was being punished for something I'd done in a past life. The lowest my hips got down to was 34.5 inches, which my French agent said was 'almost perfect' . . . but not perfect enough.

Not long after, I was dropped by my London agency for being 'too big' at a size six. To say I was gutted would be an understatement. Blame it on hunger or pure frustration, but I absolutely lost it, writing a lengthy Facebook post about the pressures of the modelling industry that by chance went viral.

To cut a long story short, an American modelling agency saw my story and flew me out to New York to sign with them. I jumped at the chance. New York felt like a new beginning for me – a chance to be free of the issues I had at home and to start something new. But while I *felt* like I was starting afresh, in reality, the worries I had about my weight still lingered in the back of my mind and I wondered if I'd ever be free of my demons.

I hadn't been out there very long when I discovered the world of plus-size modelling, full of gorgeous women working for *Vogue* and *Elle* and other huge brands. There was no denying their beauty, despite the fact they looked nothing like the models I knew (and the model I was). They owned their femininity, showcasing their rolls and thick thighs and big boobs, and I was envious: envious of women being happy with their size in a world that conspired against them.

Around the same time, and thanks to something called Instagram, I also discovered the body-positivity movement. These women weren't models necessarily (though there were a few models who supported the cause), but they were women who, like me, had been made to feel less than, simply for not fitting the mould.

To begin with, I felt like these women were being disingenuous. I mean, how could you *possibly* love your body without someone else's approval beforehand? How could you call yourself 'beautiful' if a man didn't? I genuinely didn't think it was possible to celebrate your outward appearance if you didn't look like a stick insect, yet I saw picture after picture of women showcasing their so-called 'flaws' with pride. And I wanted to experience that.

Seeing images of curvy, beautiful and – most importantly – *healthy* women began to change the wiring in my brain. It might sound silly but, up until this point, I'd never seen cellulite on another woman before. I'd certainly never seen someone actively promote their acne on social media for – *shock horror* – the *entire world to see*, or talk positively about their not-so-flat tummies. And for someone who believed these flaws should never see daylight, that felt really refreshing.

I wish I could tell you that, after seeing body-positive women both in and out of my industry, I finally learned my worth and lived happily ever after, but of course I didn't. Learning to love your body doesn't happen overnight. You don't just wake up and unlearn the countless articles, headlines and destructive imagery your brain has soaked up over the years. It was an uphill battle trying to forget everything I'd ever been taught about beauty.

As a recovering bulimic and anorexic, gaining weight was nauseatingly painful. While I admired the confidence of women celebrating their bodies, I just couldn't seem to appreciate my own body changing. But I was willing to give happiness a chance. I sat down and marvelled at the intricacy of a doughnut; I savoured meals out with girls I'd met and the stories we shared. Food was no longer an enemy but something to celebrate.

The weight soon piled on, as it does with most anorexics. And when my model agency suggested I join the curve division, I had to admit defeat – learn to let go of control and everything I'd known to be the 'right' way of being a woman.

Soon, my curves became my most favourite element about me. Apparently, though, the body-positivity movement came with its own set of rules. As I started to take photos of my body and celebrate my new-found curves, new critiques appeared. *What would you know about hating your body? You can't be body positive if you're a size ten.* And so, just as I'd started to feel beautiful, or part of a group that 'understood' me, I felt abandoned again.

I wanted to be happy, and I thought the body-positivity movement was the way in which I finally would reach inner peace. But if I looked too 'small' in photos and made a comment about how I loved my shape, the negativity online would roll in. So, in order to fit in, I started to position my body in Instagram photos to look 'bigger'. I posted photos of my tummy and my cellulite in unflattering angles, all in a desperate attempt to be accepted. The compliments on social media came flooding in.

I made other women feel comfortable when I wasn't sure I felt comfortable myself. It felt strange that, although body positivity celebrates women for their flaws, they're still celebrating outer beauty.

And then, one day, I'd had enough.

If you want to know how I truly learned to love myself, it is this: I accepted myself.

I accepted the fact that I'm never going to be 'body positive' all of the time. I accepted the fact that I'm never going to be the most beautiful person on the planet, or the most feminine, or the thinnest or the most curvy. I accepted that not being perfect is okay.

Whenever I took a break from positioning my body into appearing bigger, or stopped caring about what other people thought both online and offline, a feeling began to well up in me. *Love.* Love for myself, in a body I'd abused and hated my entire life but that had never given up on me.

The only way to love your body is to change your mind. And what I've learned is this: you won't find beauty in a dress size, or by gaining weight, or by trying to please other people, or by worrying how you come across to them. I've truly never felt better about myself than when I don't think about my body, in any way.

Happiness can come in lots of ways, like running yourself a bath, or telling yourself how grateful you are for something, or by drinking a cup of tea by a fire. There's a reason Audrey Hepburn said, 'Happy girls are the prettiest.' These little things are what light me up, and I have never felt more beautiful than I do now.

Reader, there is a happy ending to my story, and I end it on precisely that: *happiness*. Yes, I found it. I found happiness when I stopped putting so much emphasis on my outer appearance and started to consider what inner beauty meant instead.

Beauty, I've discovered, is a *feeling*. And only you have the ability to reach it.

AVA WELSING-KITCHER

on how hair and identity are intertwined for women of colour

Since joining *Stylist* as junior beauty writer, Ava Welsing-Kitcher has made representation of women of colour her priority, from helping black female beauty entrepreneurs gain more recognition to ensuring that *Stylist*'s beauty pages continue to showcase products and treatments for everyone. Her investigative article on her personal struggle with trichotillomania (hair-pulling) and others' with dermatillomania (skin-picking) had an overwhelming response from readers.

From as far back as I can remember, I've always been fascinated by hair. Its physical structure, how it pushes out from under our skin and covers our entire bodies, the fact that it's dead but many of us treasure it more preciously than we do our own skin, and that entire creative empires have been built on manipulating and capitalizing on it. The list goes on and on. Hair is more than an industry, it is intrinsically tied up in human identity – even more so if you're a person (particularly a woman) of colour. For many non-white people, 'good hair' is indicative of how society sees you and whether it accepts you. It has the power to flick that switch in your manager's mind to make them favour one colleague over another for a promotion; it can decide whether a child gets bullied at school; it can be the final twist of fate that gets somebody killed.

Since colonialism and the transatlantic slave trade, and the leaking of Western culture into places on the opposite side of the planet, the general consensus of the past few centuries has been that the closer one's proximity to whiteness, the more socially accepted you are. Despite this being common knowledge, there are still countless individuals who refuse to even entertain the idea that it could be true. We're currently approaching the edge of something that looks and sounds a lot like a 'race war' (as certain media outlets label it) but feels like the deep rumble of whole loads of people who have had enough.

Thanks to the open declaration of love for blackness within the black community observed across social media, the arts, awards ceremonies, magazine covers and podcasts, the essence of what has been deemed attractive is evolving, despite the idea of whiteness staying largely fixed. One of the most poignant anchors of love and change lies in the natural-hair movement and the resulting open celebration of Afro, textured, curly and coiled hair. Women and men across the world create content, instigate discussions and care for their own and others' hair while lifting the heritage of it high with a respect and creativity that underlines the power which so many have tried to suppress.

The importance we place on our hair has occupied my thoughts a lot recently. Yes, I've always been conscious of my hair and how it looks, as much as anyone else might be. But as I've grown out of adolescence and into my twenties, I've started to question why what's on my head holds so much power, and the leading answer is always ethnicity.

As a mixed Ghanaian and white young woman, my hair and how it looks in comparison to others' has dominated much of my life. In primary school, I knew it was different to my white friends' hair, I knew it was different to my Asian and black friends' hair, and I knew Lauryn Hill's Afro would never be mine. My mum kept my hair in two plaits, so I never really knew what its real texture was or that it could form ringlets. I moved from multicultural London to a homogeneous seaside town at the age of nine, plucked up and plopped down into an almost completely white environment, which I wasn't prepared for. Being different laid me bare for people to feast upon that uniqueness. Kids at school, along with wider society, fed me the message that straight hair was the best and that, if yours didn't grow that way, then you should do everything you could to make it look like it did. My best friend's mum called me a cavewoman after I took out my plaits and let her daughter brush my hair, and my ten-year-old self internalized the shame of being stereotyped as unkempt and uncivilized. Boys told me I was prettier with straight hair (they still do) and would feign disgust if their hand got tangled in my curls while they were trying to yank them.

A couple of years later, a bored white hairdresser lopped off over a foot of my hair (when all I had asked for was a trim) and I was suddenly confronted with another issue. To me, back then, length was even more important than straightness in portraying femininity. The hairdresser saw how quietly horrified I was to see all that hair coming off, so he straightened the curls bouncing around my earlobes. 'There, now you don't look like a little mushroom boy,' he laughed, and so began the vicious cycle. Overnight, my hairdryer, round brush and straighteners became more precious to me than my passport. I stayed up late every Sunday night to flatten my hair into submission, waking up before my family to iron out offending coils and baby hairs and avoiding swimming pools at all costs. I carried a beanie hat or a scarf with me everywhere I went, sheltering my hair before the third raindrop would even hit. I told myself that this was temporary, until my hair grew back to its old length, but I didn't realize that the excessive heat damage was causing my strands to snap along their shaft, gradually making my hair shorter.

My despairing mother, who wore her Afro hair in a variety of natural and manipulated styles, took me to her hairdresser to try and teach me how to take care of my curly hair. I hadn't seen my natural hair in four years, but Isis, a knowledgeable trans woman who had put her hair through ridiculous heat and chemical damage herself, managed to coax it back to life. She made me promise to cut down on straightening and to give my hair a chance, or it would fall out. Even though everyone loved and fussed over my hair the next day, I felt like a little mushroom again, devoid of femininity and missing the feeling of hair hitting my shoulders.

It's a common misconception that textured hair doesn't grow. When stretched out, the true length is obvious, but in its natural state it gives the illusion of being shorter. A stroll through any high-street beauty shop, be it Boots or Pak's, will show you just how much money is made from and spent on hair-growth serums, creams and oils. I had internalized the feeling that short hair was something to be fixed, and I would hold my breath and clamp those straighteners down until it was long again. All the female celebrities I looked up to with curly or Afro hair wore it straight or long, most likely with extra wefts or weaves sewn in: I took note as Leona Lewis's shiny spirals got looser and looser with every week's episode of *The X Factor*. No magazine taught me how to nurture my hair, and the people who could have told me hadn't found their voices online yet.

Everything changed when I decided on a whim to type 'how to have nice curly hair' into the YouTube search box. A few video tutorials showed me that I had been doing it all wrong. I soaked up knowledge on deep conditioning, silk pillowcases, diffusers and the dangers of sulphates. I learned how to 'stretch' my curls without heat so they'd appear longer (although, in hindsight, sleeping with Greek yoghurt in my hair and failing to rinse it out properly was not worth that extra centimetre), how to use twists and plaits to temporarily reset my curl pattern. I had stumbled upon the natural-hair movement, and it was the first place – other than India Arie's 'I am Not My Hair' music video – where I'd seen women of colour openly celebrating and nurturing their hair en masse. Blogs and forums educated me further; I avidly learned how to decipher an ingredients list and how to drop a strand in a glass of water to test how porous (read: damaged) my hair was.

My curls started to look better as time went by, but it still took five years for me to stop straightening once a month (and for nights out and job interviews, because sleek strands made me feel polished, sexier and more mature) and quit heat completely. I aimed to get to six months, then I'd reward myself with an expensive salon blow-dry. Later, as I sat in the stylist's chair and watched my hair transform back to how I liked it best, something powerful occurred: I was disappointed. My straight hair bored me; it hung predictably and obediently, it lost its shine and I missed the springy, unruly curls that had been growing bolder and braver now that I was not attacking them with heat. I couldn't wait to wash my hair.

This phenomenon is common in the natural-hair community. People document their journey to loving their hair more and more as they quit heat and relaxers to the point where they don't really want to straighten it. Photos of my hair over the years showed me that it had evolved, but my new-found love for it didn't lie in how much healthier it looked. My view of myself and my hair as an emblem of

MY STRAIGHT HAIR BORED ME; IT HUNG PREDICTABLY AND OBEDIENTLY, IT LOST ITS SHINE AND I MISSED THE SPRINGY, UNRULY CURLS THAT HAD BEEN GROWING BOLDER AND BRAVER NOW THAT I WAS NOT ATTACKING THEM WITH HEAT. I COULDN'T WAIT TO WASH MY HAIR.

my ethnicity had shifted. By letting my hair be, I felt I was accepting who I was as a woman of colour. I was tapping into my internal power, and the strength this gave me cannot be overstated. The link between hair and identity had never rung clearer; the rewards of not giving a shit about how the world perceived me, whether my hair looked unkempt one day and perfect the next, or how men preferred it, leaked out into multiple facets of my existence and made me realize that, although I'd viewed myself as progressive and 'woke', I'd also been living a life of internalized racism and sexism.

As a sufferer of trichotillomania (a body-focused repetitive disorder which makes me compulsively pull out single strands of hair), this revelation was solidified when I understood that, when my hair was curly, I didn't want to pull it as much. I first pulled out my hair the day after it was cut and straightened when I was twelve; coarse, wiry hairs stood up from my parting and an insensitive comment from a friend about them made me feel like I had to fix a problem. Trichotillomania had dominated much of my life and mental health for over a decade but, once I accepted my hair in its natural state, my issues with control and perfectionism started to fade. My mother had once commented that I was tugging out the hairs that were closer to Afro-textured hair than Caucasian hair, that I was 'denying my heritage' by eradicating them. Although this wasn't what compelled me to pull – if most of my hair had been made up of thicker strands, they wouldn't stand out as much and I probably wouldn't search for them to yank out – there was some truth to what she said, in that I had been conditioned to view kinky-textured hairs as an anomaly on my head. Against the uniform backdrop of ruler-straight hairs, the ones that wouldn't obey and stuck up no matter how many times I tried to temper them into submission were the ones that were getting in the way of having perfectly straight hair. Against a head of textured curls and frizz, these thick hairs became allies in lending body and movement and my urge to pull them decreased in a way I could never have anticipated.

Hair may just be hair to some, but to many of us it's a journey that spans across every facet of our lives and dictates how we position ourselves both in society and in our minds. Finding a community online helped me feel settled in my identity. The natural-hair movement does have the drawback of using light-skinned women like myself and those with looser curl patterns as its poster girls, while ignoring those with dark skin and Afro hair who started the movement in the first place. Yet I feel that the undercurrent of celebration, creativity and acceptance which runs through the natural-hair movement's veins has the strength to tackle issues like colourism and appropriation head on. The world is waking up to how vibrant minority ethnic culture is and everyone wants a piece of it for themselves; they can try and 'cash-crop cornrows', as Amandla Stenberg described it in her powerful video on the appropriation of black culture, but our hair has always and will always be ours.

JESS GLYNNE

on acne and self-confidence

North Londoner Jess Glynne's first album, *I Cry When I Laugh,* debuted at number one in 2014. It has since been certified triple platinum, has spawned 12 million singles sales worldwide, with thirty-nine weeks in the UK Top Ten and a sold-out UK arena tour. Her second album, *Always in Between,* was released at the end of 2018 and promptly nominated for two Brit Awards. Jess's independent spirit has made her name; she sings about standing on your own two feet in an empowering mix of pop, soul, R&B and house music.

When I was growing up, the process of beauty and make-up was fun. I would watch my mum intently as she got ready to go out. I remember being fascinated by the way she applied her eye make-up and lipstick. To me, she was the most beautiful and glamorous woman in the world. She worked in the music industry, and she and my dad were – and still are – very sociable. They loved getting glammed up and going out. That was a huge influence on me as a child, because I wanted to be just like my mum.

In my early teens, my biggest beauty inspiration became my older sister, Rachael. She's two years older than me, and I thought everything she did was so cool. I would imitate the way she did her eye make-up, with loads of black MAC Eye Kohl eyeliner and mascara. That's how I learned to do make-up, from watching her. I loved playing with different beauty products when I was a teenager, as it felt like a fun way of extending your personality, just for a while. Like trying on different identities.

I've always been around music and was inspired to write and sing from a young age. When I signed my recording contract in 2013, everything happened so instantly and my life went a bit crazy. It was amazing, of course, but touring for months on end meant working long hours, and a day off was a rarity. At first, because I was just starting out and was still so young, I didn't know enough about what was best for me and I didn't have the confidence to ask for it, so I worked with make-up artists who would layer so much foundation on my face. Looking back, it was mad that they would put that much make-up on me, when I was young and I'd always had good skin. Eventually, my skin erupted into terrible acne. I was twenty-four and had never had any issues with my skin, but now I was having more and more make-up being applied to cover my acne, and it was only getting worse underneath it all. I realized that my skin is actually very sensitive and couldn't handle being overloaded with clogging foundation on top of a stressful and exhausting schedule.

Acne is a horrible thing to live with, because it affects your confidence so much. It might sound shallow to say that something like how you look can affect how you feel, but it really does. It's right there on your face, and you're aware of it all day, so of course it makes you extremely anxious and wrecks your self-esteem. I never felt that I looked like a pop star anyway, so it was just another thing that made me feel insecure. In this business, you are being judged all the time. I have never had an issue with how I look, but neither have I ever thought that I am the traditional idea of a pop star (whatever that is!). Consequently, being catapulted into the limelight made me focus a lot on my appearance. I spent so much time researching products and skin specialists that might be able to help me, but nothing seemed to work.

It wasn't only the heavy make-up that was having a damaging effect on my skin. I believe the acne was a direct result of the amount of stress I was under at the time. I was constantly working, constantly anxious and constantly exhausted – so of course that would show on my skin. No break from work meant no break from make-up. It was a vicious circle, because I knew it was making my skin worse but I had to wear it to cover my spots.

Make-up had always been something that I enjoyed, a fun part of life. But now, for me, it had become something negative, both the reason for my bad skin and the disguise for it. I didn't get any joy out of having foundation caked over my bad skin by a make-up artist who didn't understand me.

In 2015 I got my wake-up call. I had a polyp on my larynx and it was haemorrhaging. Going in for surgery was terrifying, and it's still the worst thing I've ever had to go through. I couldn't even speak, let alone sing, for three weeks – and all with the anxiety of not knowing if my voice would ever fully recover. I had to cancel a tour with John Legend and a performance at Glastonbury, which was heartbreaking, even more so because I had brought this on myself by working too hard. I had been overdoing it, not only with singing every day but also with all the talking you have to do as part of the promotional work as a musician. The doctor asked to see my schedule and was horrified to see that I had no rest time and absolutely no days off.

It was such a difficult period, but it changed my life. Having enforced time off brought the epiphany I needed to make some big changes. Having time to rest – make-up-free, of course – meant that my skin cleared up, as did my anxiety. Now I knew what I had to do: I had to learn that sometimes I have to say no. It was a tough lesson, particularly for someone who wants to grab every opportunity. But I can't do everything, I just can't. I also learned that, if I want a day off, I have to make it happen – no one is going to offer it to me. Luckily, my vocal chords completely recovered from the surgery, but that operation, and time away from the whirlwind of my career, changed everything. Realizing that my lifestyle is my own responsibility was such a turning point, because I had just been doing what I was told. After that, it dawned on me how important it is to have control of my life, not only over my work schedule but also over my health and my appearance.

Now I work with a great make-up artist and my acne days are over. I don't wear foundation any more, just a nice light BB cream. And I have completely changed my diet, which used to be quite unhealthy, and I really try and look after myself in terms of nutrition and exercise. That makes such a big difference, not only in terms of my skin but also in terms of my mood and energy levels. I still sometimes get breakouts if I'm stressed or tired, but now I hate to cover it up and I really try and be honest about it, particularly on social media.

The mental health of young people is a massive issue at the moment, and the feeling many have that they're not living up to the image of perfection so often portrayed on social media is a big part of that problem. So many celebrities only post pictures of themselves that are beautifully lit, taken after hours in Make-up, and then altered and filtered until they look perfect. I don't judge anyone who wants to spend a lot of time and money on their appearance, but I want to put myself forward in a more authentic way.

When I was younger, the celebrities I looked up to for beauty inspiration were models like Kate Moss, whose look was very simple and natural. And, if you look at the Spice Girls, who were huge in the nineties, when I was a kid, they had fun with make-up but it wasn't the primary thing that I noticed about them. There was no contouring, no fillers, no filters. Now there is so much artifice, the images projected are far more unattainable.

Nobody is perfect. People wake up and have scruffy hair and spots on their face, and that is all a part of life. It's important for someone with a platform to say that. That's why I want to talk openly about my acne. I am aware that I have a responsibility to think of the mental health of girls who listen to my music or follow me on Instagram. In that way, my bad skin was actually a good thing, because I was able to be really honest and show the real me. I wrote the song 'Thursday' with Ed Sheeran because I wanted to explore these insecurities. This song conveys the idea that by being honest with who you are and stripping everything back, you can be the best, most beautiful version of yourself.

NOBODY IS PERFECT. PEOPLE WAKE UP AND HAVE SCRUFFY HAIR AND SPOTS ON THEIR FACE, AND THAT IS ALL A PART OF LIFE. IT'S IMPORTANT FOR SOMEONE WITH A PLATFORM TO SAY THAT.

The lyrics to the first verse are:

I won't wear make-up on Thursday,
I'm sick of covering up,
I'm tired of feeling so broken,
I'm tired of falling in love.
Sometimes I'm shy and I'm anxious,
Sometimes I'm down on my knees,
Sometimes I try to embrace my insecurities,
So I won't wear make-up on Thursday,
Cos who I am is enough.

I feel that owning up to my insecurities, anxieties and self-doubt at times is really important. Being honest is something I live by and, hopefully, letting my fans know that they are not alone in their insecurities will inspire them to accept who they are, as they are, and be capable of standing up for themselves.

Now, I'm no longer afraid to speak up and ask for what I believe in. I think that's an important example to set for young girls who might look up to me: don't worry about being labelled as 'difficult' and don't let anyone else try to steer you away from what you know is best for you. You have to be strong, and I believe I am; it's something my parents instilled in me. I have never been afraid to work, and I still work hard, but now I know that it's important to take time out, too, for the sake of my physical and mental health. Ultimately, I'm so lucky to be doing my dream job. In 2018 I became the first British female solo artist to have seven number-one singles in the UK singles chart, which is mad. I really do appreciate how amazing it is that I get to do this. However, even when you do the job of your dreams, you can't be happy when you overwork yourself and never take time out to appreciate everything you have achieved.

The last few years have really been a journey of self-acceptance, and it's funny that having acne and learning how to feel like myself through beauty was what brought me to this point. Life is too short to spend it doing only what others expect of you. As women, we have to stand up for ourselves and, if I can inspire young girls and women to do so through my music, then I will be very happy.

HARNAAM KAUR

on defying beauty ideals

Otherwise known as the Bearded Dame, Harnaam Kaur is a body-positivity warrior, model, world-record holder and activist. Born and raised in England, Harnaam suffers from polycystic ovary syndrome and spent her youth removing her unwanted body hair to avoid taunts from her peers. Driven to self-harm, she decided at the age of sixteen to embrace her body in all its hirsute glory. Now, Harnaam uses social media and public-speaking engagements to spread her message of diversity and self-acceptance.

I was sixteen years old when I decided to grow my facial hair. I made the decision after my GSCEs and let it grow out over the six-week period of the summer holidays. So when I returned to the same school (in Slough) for the sixth form, I was a girl with a beard.

It took several years for me to reach the point where I felt able to do this, and it was bullies that had pushed me to that point. Indian people are quite hirsute anyway, and I have PCOS (polycystic ovary syndrome), which is a hormonal disorder affecting between 8 and 20 per cent of women worldwide. The condition affects how your ovaries work and the side effects can range from irregular periods and infertility to excess androgen – high levels of 'male' hormones in your body. This can mean that not only do you lose hair from your head, you also might have excess facial or body hair.

I didn't actually realize that I had facial hair until I started being tormented for it. The bullying started in primary school – just casual taunts at first, like 'Harnaam's got a moustache.' I was in Year Six when I remember going into the bathroom, looking in the mirror, and thinking, *Wow, I really do have 'unwanted' hair.*

The bullying stepped up in secondary school. In my early teens, it was a case of just trying to get through the day, keep my mental health intact and not get beaten up. I was pushed against lockers, I was cornered and had balls kicked at me, I was stabbed in the hand with a pen . . . I even received death threats. They said they were going to burn my house down while I was sleeping. I suffered fat shaming as well, including having food thrown at me; weight gain can also be a symptom of PCOS.

These experiences led me to have panic and anxiety attacks in school. It was around then that I started self-harming. I hated my body and I wanted to punish it because my body was the reason that I was being bullied. It was also a way of releasing energy and trying to gain control. People who have not been through self-harm might find it hard to understand, but I just wanted to feel.

At my lowest point, I felt suicidal. I actually felt that the world would be better off without me in it. Then I had an epiphany: how dare I allow myself to feel like this when my bullies are happily going out and having fun with their friends? I reached a state where I felt like I'd been through everything. I'd heard every negative name you can imagine. There was nothing that anyone can say any more that was going to shock me. I had hit rock bottom and there was nowhere else for me to go. I had tried so hard to remove my beard every day and I was still bullied so I decided to try embracing my natural body. I thought, *Well, if you're going to bully me, I'm going to give you something to bully me about.*

It sounds crazy that I reached that conclusion overnight, but it was desperation. Of course, my family were against the idea initially. My parents feared that I wouldn't be able to lead a 'normal' life; as a bearded lady, I wouldn't get married or find a job. But I couldn't continue shaving, waxing, plucking, threading, using hair-removal creams and still be tormented for my facial hair. I didn't have a goal in mind at this point. I just knew that, if I wasn't going to end my life, I was going to have to grasp hold of it and live in an authentic way.

I remember the first day I went back to school after that summer. Everyone was looking at me. Everyone was staring. I was used to bullying, but when you've got the whole school staring at you and taunting you, it was a very difficult experience. I realized that, if I was going to be different, then I had better get used to abuse. And that's what I had to do.

Opening an Instagram account didn't feel important at first. Initially, I barely used it, and I certainly had no idea where it could take me. It was only once I started using it to talk about my journey and my story that it suddenly got really big for me. In 2014 things escalated very quickly and I started to receive offers to be profiled in the media and appear in modelling campaigns.

Now I try to use it as a platform to help people. I know that I come across as very confident, but I want to show the vulnerable side of me, so I talk about how I have been able to overcome difficult times in my life.

Of course, there is a negative side to social media as well, with filters and Photoshop and people changing their bodies to look a certain way, and that can be very damaging to the mental health of people who are insecure about themselves. You have to be very mindful about the type of people you're following. Our eyes consume images, and the more we see a certain type of image, the more we want to attain that particular beauty ideal. That's why representation matters. That's why, being this different, I try and be as visible as I can.

I know there are other women out there who are able to grow facial hair and that they feel uncomfortable about it. So I'm here to show people that you can be different and still achieve everything you want to. And I'm here to show you how to do it.

It's amazing that people who feel alone can find communities on social media, platforms that are rooting for a better way of living. Being able to interact with them or connect with them is great. Having said that, so many people feel they can open up to me about their issues and I do find it hard not being able to help as much as I want to. People send me direct messages telling me they're going through a really hard time, and often they're thousands of miles away. It breaks my heart that I can't help them. I used to be really hard on myself and would be up until 5 a.m. emailing someone in Texas because they're struggling with their mental health. But I've realized that I can't be everyone's saviour and that my welfare is also important. I'm not as hard on myself as I used to be.

I still have bad mental health days. I'm still ostracized. I'm still discriminated against. Bullying is still something I face. People might think it's easier now that I'm in the public eye, but they don't see the amount of strength it takes me just to get out of the house every morning. There is so much injustice out there, and we live in a shallow world that will criticize you for being your authentic self. But people go through hardships and horrendous times in their lives and then become able to help other people. For a long time, I'd ask myself, 'Why is this happening to me?' Then I realized it's because I have a story to share with others to help them with what they are going through. I feel really lucky to have the opportunity to speak about my story and use it to help others. Starting to speak publicly about overcoming negative self-image and becoming a body-positivity activist happened so naturally. I started doing this work when I was about twenty-four, and I'm so grateful to have found my purpose in life at such a young age.

What matters to me is that I accept myself. But sadly, receiving abuse is still my norm. The world is still not inclusive. Even in the UK, women don't have equal pay. So how can I expect people to accept me as a bearded lady? If you're not giving cisgendered heterosexual women equal rights, how can I expect people to accept me when I'm a woman of colour with a beard and a turban?

My turban is my crown, it's part of my identity. Royals wore turbans back in India and Pakistan. It's a very powerful piece of clothing. And I call my beard 'she', rather than 'it', because she deserves to be celebrated. She's a part of me. Why would I disrespect her? She's an extension of who I am. So I decided to give her a name, Sundri, which means 'beauty' or 'beautiful' in Sanskrit. Society will label

MY TURBAN IS MY CROWN, IT'S PART OF MY IDENTITY.

me with the most horrendous names, so it's up to me to think about my body in a positive way.

Now, people can deal with me being a bearded lady because I own it. The reason why so many people have trouble accepting their bodies is because we're used to treating our bodies in the most negative way. We see, on the TV, in advertisements, on billboards, 'Are you beach body ready?' I couldn't give a crap about all of that. I live by 'My body, my rules.' Beauty standards are not going to dictate how I should live.

It has taken many years for me to be this confident and strong within myself. I've had to fight many demons. My advice to anyone struggling with self-acceptance, in any form, is that it's actually okay to struggle with self-love. It's not something that you should feel guilty about, or ashamed of. How can someone who's gone through years and years of abuse suddenly turn around and say, 'I love myself now'? It takes a lot of personal growth to be able to come to this point, and I struggled with that for years.

The first step towards striving to be happy within the body that you have is authentically saying to your body, 'I'm sorry for treating you badly, but I promise you that I will be better.' And then you can start the process of healing from any abuse, bullying or body shaming that you may have gone through. Because it is through doing this that you will be able to find the freedom to say, 'This is who I am, and I'm going to fucking own it.'

When people ask me if I'm always going to wear my beard or if I'll remove it one day, I find it so annoying. Because, to me, that's like asking, 'Are you always going to have that hand on your arm, or are you going to remove it one day?' Or 'Are you going to remove your eyeball, or do you think you'll keep it?' It feels really weird for me to answer that question because why have I gone through so much in my life, from such a young age, to decide to change myself now? My beard is here for a reason. It's my identity.

Despite everything, I'm grateful for all the abuse and bad times I've been through, because that is what made me who I am now. It has allowed me to be an inspiration for others, to show them that you can come out the other side of hard times. Finally, I feel free.

CAITLIN MORAN

on her larger-than-life beauty identity

The eldest of eight children, home-educated on a council estate in Wolverhampton, Caitlin Moran published a children's novel, *The Chronicles of Narmo,* at the age of sixteen and became a columnist for *The Times* at eighteen. She has gone on to be named Columnist of the Year six times. Her multi-award-winning bestseller, *How to be a Woman,* has been published in twenty-eight countries. Her two volumes of collected journalism, *Moranthology* and *Moranifesto,* were *Sunday Times* bestsellers, and her novel *How to Build a Girl* is currently being adapted as a movie.

Growing up, we were too poor to buy make-up. But we did have felt tips, so consequently my first attempts at putting on eyeliner were with a black pen. It would bring you up in a massive scaly rash, which we described as 'the mermaid look'.

At thirteen, I realized the only beautiful thing you can do if you have no money is to grow your hair. That epiphany resulted in the first big transformation of my life – I stole a comb from the local chemist and started back-combing my hair. I thought at the time that if my head looked big, then my body would look smaller in comparison. And I could make it as big as I want, it could be as big as my imagination.

My sisters and I got our beauty tips from the library. We were very nineteenth century in our beauty regimes then – we'd read books on how to create cosmetics using household ingredients. It would involve things like putting egg on your hair to make it shiny, but then not getting the temperature of the rinse right and ending up with scrambled egg in your hair.

One of my biggest beauty breakthroughs was realizing you shouldn't wear nappies on your head. I read that in the eighteenth century women used to use pads, called 'rats', to make their hair look bigger. The problem was that the only thing I had to use as a rat were my younger siblings' terry towelling nappies. So I'd basically pile a nappy on my head and arrange my hair over it, which was fine until I hit a brisk wind. Given that my primary hobby at this point was being chased down the street by yobs who would run after me throwing gravel and shouting, 'You fat lesbian!', having a nappy on my head stepped this up to an intolerable degree of bullying. It was at this point that I realized I needed to leave Wolverhampton, so I started to write a novel. In a way, my beauty failures gave me the ambition I needed to leave.

The first proper perfume I ever bought was Chanel No. 5. After I moved to London, I heard people talking a lot about 'signature perfumes', so I thought that was something I needed. Before, I'd used the same as every girl in the nineties, namely Body Shop's Dewberry perfume, or their White Musk, if you fancied a more sophisticated smell. But then I interviewed Richey Edwards of the Manic Street Preachers, who informed me that Silk Cuts were 'the cigarettes of the working-class girl' – which I started smoking – and that the best thing to cover up the smell was Chanel No. 5.

I always want my beauty look to reflect me. In the past, whenever I did a shoot I'd be given a very straightforward, glamorous look, which would make me feel a bit sad. It didn't communicate anything and seemed a bit bland. It was like I was pretending to be some glamorous woman, with maids and a massive house and loads of Lululemon leggings. That's not who I am. It made me feel uncomfortable to be communicating the wrong thing. I feel like part of my job is a responsibility to look cheerful and comfortable and myself – you just don't see enough people looking like that.

Big eyeliner is definitely my go-to look – three inches of eyeliner paired with a bright eyeshadow. I love a bold emerald green or peacock blue – to have a clown eye is a wonderful thing. It's such a low-maintenance look as well, because it takes thirty seconds and you can do it on a bus while drunk, as indeed I have done.

My beauty icon is the puffin. It's hard to explain that to make-up artists, but some people look better trying to look like an animal than someone off *Love Island*. I'm a round bird, I primarily dress in black and white, I wear bright eyeshadow and lots of black eyeliner. I have to explain that my eyeliner isn't meant to look sexy or glamorous or feline or 'flicky', it's supposed to look like a puffin. That's how I want to look.

People ask whether it's anti-feminist to wear make-up, but I say, if David Bowie wore make-up, then, as a feminist, I can also wear make-up. There's a lot more gender fluidity going on in terms of make-up today. If men can wear it, then why can't women, too? My argument with feminism is that you can tell sexism is happening when it's something men are doing but women aren't or, conversely, if it's something women worry about but men don't. No one has ever said to Bowie, 'Why are you wearing make-up?' Everyone just loved Ziggy Stardust. Maybe we want to look like Ziggy, too.

My white hair streak is my calling card. The white grows naturally, but I used to bleach it to accentuate it. The problem was, it would only look good for a couple of days before it started looking yellow and shit. So, now, I have a few hair swatches I simply glue to my hair when

I want it on show. People associate it with me now so, if I don't want to be recognized, I don't put it in. It's like my open/closed sign – if the white streak is in, I'm open for business. If not, I'm off duty.

There are millions of reasons why I love make-up. It's not just about looking glamorous or sexy. If that's what you want, then go for it, but for me it's about making you happy, making you fabulous and giving you joy. The truth is you very rarely need to be hot and sexy. Walking down the road, for example, do you need to be hot and sexy? No, you're on the way to M&S. You're not going to bang someone now. So why bother?

Make-up is the great equalizer. If you're not born typically 'beautiful', you can learn a skill and apply make-up to look however you want. You can change how you define yourself, and that is magic. Anything where a woman can learn a skill and change her self-worth is a magical thing.

MAKE-UP IS THE GREAT EQUALIZER. IF YOU'RE NOT BORN TYPICALLY 'BEAUTIFUL', YOU CAN LEARN A SKILL AND APPLY MAKE-UP TO LOOK HOWEVER YOU WANT. YOU CAN CHANGE HOW YOU DEFINE YOURSELF, AND THAT IS MAGIC. ANYTHING WHERE A WOMAN CAN LEARN A SKILL AND CHANGE HER SELF-WORTH IS A MAGICAL THING.

It's not anti-feminist to say a product is anti-ageing, it's just stupid. They might be able to plump you out for a bit, but they don't stop you ageing. Whenever I see a twenty-year-old trying to flog me one, I just turn the page. There's a widening of the kind of women we see in the media, but it's still only a fraction of the true demographic. Until there is a female equivalent of Seth Rogen – and I say this with love, as he is the kind of guy I would bang – who looks like a sofa with the stuffing coming out of it but gets million-dollar film deals regardless, we know we're not anywhere near true equality.

Big hair and big eyeliner have always been the constant. The changes have come more below the chin, with a variety of successive breakthroughs where I'm finally happy with my body.

I grew up loathing my body. My biggest dream as a teenager was that I would be caught up in a massive car crash and the miracle that is the NHS would rebuild my body while removing about four stone in the process. That's an extreme level of self-loathing for a teenager. Then, when I gave birth to my first child, it was very traumatic and suddenly I felt sympathy for my body for the first time. I'd seen it entirely as a problem before, but then when I gave birth I felt very sorry for my body. It was like, 'Aw, mate! That fucking chafed!'

Thinness bores me now. We're in such a sculpted-body age. Look at Kim Kardashian – her body is a feat of engineering and cash. Unless it's your full-time job, you will never look like that. There's no traction to it either. Twenty years ago, nudity was a big deal, but now you see naked tits all the time. What's going to catch your eye these days is an imperfect body. That seems rarer and more interesting to me. We've seen enough physical excellence now; the pendulum is starting to swing the other way. We want realness.

Yoga has definitely improved my body image. It teaches you that you can make yourself happy by essentially just massaging yourself for an hour. I don't do yoga classes, I just follow videos by this Texan stoner girl called Adriene. Her mantra is 'Just do what feels great.' It sounds like an obvious thing to say, but it's actually revolutionary. How often are you told as a woman to just sit down and do something which is free, makes you feel fucking great and is purely for your own pleasure? Not to tighten up your pelvic-floor muscles to become better in bed or to fit into special jeans. Just because it feels good. Rarely.

My attitude to body hair is 'Live and let die.' I remove it from the places where it might cause terror rather than offence in public, just the edge of the bear's head. I'm an incredibly lazy person. Unless someone is knocking on my door saying, 'Dude, sort your minge out,' I just let it be.

STELLA CREASY MP

on the politics of beauty for women in public life

Stella Creasy has served as the Labour and Co-operative Member of Parliament for Walthamstow since 2010. While in Parliament, she has campaigned on issues such as payday lending, access to abortion, women's rights and PFI (private finance initiatives). Before becoming an MP, Stella was the Head of Campaigns at the Scout Association, working to support young people across the country in developing their advocacy skills. She has a PhD in social psychology from the London School of Economics.

If we want a revolution, as 'ladies', the first question is, naturally: should I wash my hair first?

All too often for female politicians, it's our appearance, not our arguments, that gets the primary attention. Beauty may be in the eye of the beholder but, in our unequal world, it is also in the eye of the electorate. For those women who want to change the world, the expectations about what they must look like in order to be 'taken seriously' can be crippling – in the case of footwear, literally. For a real revolution, we need to rip up the old rules and champion the beauty that diversity can bring to all our lives.

I've never enjoyed being told how I should look. When I was eleven, my mother and I had a furious row about a pink bolero jacket which she felt would be perfect for me and about which I was ready to seek adoption if she made me wear it. Even then I knew the fight wasn't really about the

outfit – it was about being able to tell me what to do. How I chose to present myself – and, as a teenager, it involved, at points, blue lipstick, lumberjack shirts and pink hair dye – wasn't about style so much as the substance of my attempts at rebellion against the mainstream. I didn't want to be silent or conventional in any way. My choices, like my opinions, were about being different.
I didn't want to fit into a world that didn't fit the world I wanted to live in.

In adult life, that pressure to look a certain way in order to meet someone else's assumptions has been non-stop. The temp-agency boss with the shocking-pink lipstick who demanded I buy a blazer and lose weight when I was eighteen so she could comfortably send me to make cups of tea for bankers. The Party staffer who told me when I was picked as a candidate that I needed to have my 'colours' done – apparently, I was a 'sludge' – for people to find me appealing. I had already spent a month's salary on a frumpy pillar-box-red suit, heavy foundation and bouffant hair for the selection process because I was told to look 'older' – again, to be 'taken seriously'. I'm not sure the Dolly Parton effect it produced swung any vote. Each of these moments was a reminder that, as a woman, your value lies in what others see of you and find attractive in you, not in what you say.

That hasn't stopped, even after having won elections. As a female MP, I receive comments about my appearance on an almost daily basis, from the man who wrote asking for video footage of me pulling off and on knee-high boots I'd worn on TV to the commentators who suggest I'm wearing make-up for them rather than because I'm onscreen. When I cut my hair short, several male colleagues spent a week debating whether this meant I was now a lesbian. Because, still, when it comes to women, how you look is used to judge who you are, whatever words you say. The stereotypes are not just about your hair or your make-up – as a new MP, I was warned by a female whip to use my doctorate in my title in Parliament. Yet again, this was to help me to be 'taken seriously', because I was a young blonde woman. I firmly told her that if people wanted to use my hair colour as a guide to my opinions, they would only do it the once.

As I get older, the presumptions made about me on the basis of my appearance change, but they don't stop, making this all feel like a game you cannot win, whatever you do. Yet I have also become stronger at fighting for what I believe in – and so less willing to wear eyeliner and big hair just to conform to social expectations about how women should look if they want to be noticed in a man's world.

Yet, like over-plucked eyebrows, outdated trends dominate our discussions of what leadership

CHRISTIAN LOUBOUTIN ONCE CLAIMED, 'HIGH HEELS ARE PLEASURE WITH PAIN.' ANY WOMAN WHO HAS TRIED TO WEAR STILETTOS FOR MORE THAN HALF A DAY KNOWS THE PLEASURE IS FOR THE AUDIENCE, NOT THE AGONIZED WEARER.

looks like, to the detriment of all concerned. Even when we elect women, we still restrict them by expecting them to look, sound and style themselves in a certain way in order to be seen as 'credible'. Break those conventions and, no matter what title you have, you face an uphill struggle to be heard before you've even opened your mouth.

Take the matter of what so-called 'power dressing' means for our feet. Christian Louboutin once claimed, 'High heels are pleasure with pain.' Any woman who has tried to wear stilettos for more than half a day knows the pleasure is for the audience, not the agonized wearer. Popular culture shows women in high heels to be important, powerful or just plain worth watching. Wear a sensible shoe and you evoke a very different impression. Flat shoes, it's a flat no. When with every step you make you're wincing in pain, it's hard to focus on anything other than when you can sit

POPULAR CULTURE SHOWS WOMEN IN HIGH HEELS TO BE IMPORTANT, POWERFUL OR JUST PLAIN WORTH WATCHING. WEAR A SENSIBLE SHOE AND YOU EVOKE A VERY DIFFERENT IMPRESSION. FLAT SHOES, IT'S A FLAT NO.

down. Sensible shoes may make for the headspace to make more sensible decisions – but we have yet to see any female PM turn up in trainers to Downing Street.

The shoes they wear, the foundation they choose, the haircuts they have – for women in the public eye, in our unequal world, getting ready each day is never solely about their comfort or enjoyment. It may be 2019, but the attitudes around the physical appearance of women in public life stay firmly rooted in the 1900s. When Theresa May met with Nicola Sturgeon following the European referendum, one headline read, 'Never mind Brexit, who won legs-it?', as commentators were overcome at the sight of these two grown women leading their nation's negotiations and wearing skirts while talking business. Angela Merkel, the German Chancellor, is one of Europe's most powerful people. She has still had to put up with reams of columns written about how 'dull and frumpy' she is and what that tells us about her pragmatic sense in office. Even now, in modern Britain, it's still acceptable to suggest that the cut of a woman's trouser suit or the shape of her eyebrows tells you more about her capacity to lead than her words.

Conversely, few of my male colleagues ever seem to have their appearance analysed in this way. I've yet to hear one wail in frustration at hearing the press pontificate about whether David Cameron or Gordon Brown having grey hairs meant good or bad news for the economy. Ironically, male politicians do have their attire publicly controlled. Parliament has a strict dress code for men: suit and tie at all times. In the summer, there is great excitement if it becomes hot enough for them to be given permission to remove their jackets.

This lack of a formal code for how women should look does not mean freedom, it simply reflects how alien our presence in public life still is, even now, despite over a hundred years of some women having the vote. Shortly after being elected in 2010, a male MP demanded a public debate on whether it was acceptable for female MPs to wear denim in the chamber. This revealed not only that he had been staring at the bottoms of the women around him but also just how far our public life has yet to go. For decades, when women have spoken in the House of Commons chamber, they have been accused of using this 'freedom' to present themselves for the entertainment of men as a way of winning an argument. In 2016 a *Daily Mail* headline read, 'MPs who flaunt their, er, agendas: Feminists may howl, but there's always a reason ladies of the House parade their curves'. The article claimed that women MPs were wearing low-cut tops to 'distract' male colleagues because they knew the effect it would have, and even claimed that the women got promotions as a result. Thus, even when women do conform to the requirements of 'power dressing' – wear the lipstick, do the big hair – their voices are not necessarily heard but their motives are still questioned.

That's because telling women how to look, shaming them into conforming to a certain image, publicly judging them on their make-up, clothes or hair, is never about how a woman actually looks. Whether they choose to wear a burka, an Amy Winehouse beehive, to dye their roots or not, it's about controlling them and their place in society. The power to tell women that their worth is defined by the gaze of others is the power to tell them what matters about them. That their role is to evoke pleasure for others, whatever the pain to them. To be beautiful rather than to be equal.

This is not a call to abandon beauty, to reject make-up or shave your head. It is to say we should call out how and when conventional notions of beauty are used to close down the causes and concerns of women. Let us also celebrate the beauty we would see in a world in which women of all ages, ethnicities, shapes, sizes and abilities feel empowered to say what they think and are respected all the more for it.

Put simply, you shouldn't have to have a perfect blow-dry to be heard. If you want to wear a PVC skirt in Parliament – as I once did, to the horror of the *Sun* newspaper – this is not a sign of mental weakness, or incidental to the speech you make. For both men and women, old and young, your style should be a backdrop to, not the be all and end all of your worth. Equality for men and women will not be won on the right to appear in public with unwashed hair and face equal derision. It will be achieved when each has the right to appear how they like, without anyone using this to claim it tells them what they think.

Our world is riven with inequalities that hold everyone back – whether it's the persistence of violence against women, the gender pay gap or the domination of the male, or that we are pale and stale in our decision-making processes. The beauty of diversity still doesn't inform our society or our democracy, to the detriment of us all. Men and women both miss out on the benefits this would bring in creativity, resources and thinking. Until this changes, challenges like the global economy, Brexit and international security will continue to defeat us, because we're using only half the available brains in our country, however glamorous they are.

So please, wear whatever lipstick will help you focus on kissing the patriarchy goodbye, if it feels right for you. Let's work together to ensure that we don't let anything – misogyny, body image, the ability to apply blusher in a way that makes people listen – stop us securing that revolution and creating a more equal world. One lace-up brogue or blue eyeshadow at a time.

YRSA DALEY-WARD

on beauty as the great solution

Known for her online presence and her talks at global speaking events including TED, Yrsa Daley-Ward is a writer, poet, actor and model of Jamaican and Nigerian heritage. She was raised by her devout Seventh Day Adventist grandparents in the small town of Chorley in the north of England and now lives and works in New York. In her writing Yrsa tackles issues of social awareness such as mental health, sexuality, love, grief and addiction. She is best known for her debut collection, *bone* (2017), and her autobiographical novel, *The Terrible* (2018).

'Beauty makes people stay, I thought. *Beauty makes people listen to you. Beauty makes people fall in love with you and not know what to do with themselves.'*

YRSA DALEY-WARD, *THE TERRIBLE* (2018)

I used to see physical beauty as the antidote to anything difficult. The magical fairy dust that was bound to make even the worst things become good in the end. Being an early, avid reader, my obsession with being beautiful might very well have developed from the messages in the children's books, and films, surrounding me. Beautiful people always seemed to make it through the worst of circumstances. It was a complete fact; the proof lay everywhere. Beautiful people became rich, famous and happily content. This was especially true of women. Princesses and paupers went on to rule kingdoms if they were beautiful enough. Cinderella was practically an orphan, had an abusive step-parent and two awful siblings but was so incredible to look at that the prince took one glance at her and fell hopelessly in love. Following the matter of a party and a lost glass slipper, her life was changed and everything horrible that had happened before was forgotten – *happily ever after.*

Sleeping Beauty was *discovered* by a prince in the woods (in the film cartoon version, at least), a prince who was dreamily wandering by, with nothing better to do. He fell in love with her singing voice first . . . but laid his eyes upon her and loved her instantly. Following the matter of a spindle, a pricked finger and a non-consensual kiss (questionable), they lived *happily ever after.*

Snow White was fatherless, like me, but was so wondrously pale and beautiful that she was able to enlist seven little people to house and take care of her en route to the most coveted prize in life: a handsome, though otherwise unremarkable, prince. Therefore, yeah . . . they lived *happily ever after.*

'Beauty and the Beast' (well, the clue is in the title). She was beautiful; he wasn't. Following the matter of a curse, a kidnapping and a wilting rose, *happily ever after.* Oh, and he became beautiful again, too. She broke the spell by falling in love with him despite himself, despite his appearance. God forbid he'd remain less than.

The aforementioned were also all white, too, but that is a different essay.

No, perhaps it's this one.

Looking back, though, I am not entirely sure of the catalyst – not sure which thing came first. The need to be seen and thought of as beautiful might have also begun with my thoughts about my mother, Marcia. I was raised apart from her by my strict religious grandparents in a small village up in the north of England – which sounds rather like the beginning of a fairy tale. I was five when I started to wish and pray hard for beauty, on every birthday cake and in every prayer meeting at church. Marcia was a beauty. Whether she was absent or visiting at the weekend. Whether in person or dreamed. Whether she was delivering bad news or had a new boyfriend we hated or was so tired from her hospital night shifts that she didn't know what day it was. She had a bright smile and hair that shone and smelled like the salon. It hung to her shoulders in a curly perm (this look was highly aspirational at the time). She worked nights, so my brother and I didn't see her as much as we liked. But when we did, when she'd breeze into our lives in her new white car with her long red nails and some KFC for us, we thought she was *everything*. She really was. She'd look at me for a long time, hold my face and exclaim, 'Look at my beautiful, beautiful one.'

But I didn't feel it at all. I was completely convinced otherwise. Perhaps my mother just said that to make me feel better about the fact that we couldn't live with her. Or maybe it was because my dad and my brother's dad were absent – one lived a few miles away and didn't want to know their child and the other was in another country with his other family.

I didn't believe in my own beauty and so I couldn't wait to look like a thing that people liked. I had the sense that it might make me loveable. I had an idea that it might make me more accepted and would, at long last, make people want to stick around.

My little brother and I lived with Grandma and Granddad in Chorley, Lancashire. Growing up in this town, there was only one other family of colour that I knew of. I was the only black person in my primary school; one of three when I arrived at secondary school; one of four when my brother arrived some years later. I felt like a different thing entirely. 'It's not your fault you're coloured,' well-meaning friends would say. My hair grew up and out, not down. In class, I was two heads taller than everyone else and filling out everywhere tremendously, the first person to wear a bra, at least a year and a half before everyone else. I thought I needed to shrink. I desperately longed to *get thin* and look like everyone else or, at the very least, not like me.

In private, I began to make 'beauty lists'.

- *Plan one. Ask Grandma about hair relaxer. Or a perm. (The white girls used Pantene Pro-V in their hair. I needed to get hold of that and mix it with our horrible shampoo to make my hair come out straighter, shinier, LONGER.)*
- *Plan two. Do exercises every day from Grandma's yoga book and DO NOT EAT TOO MUCH. That means NO CRISPS, NO DESSERTS, only GRANDDAD'S DIABETIC BISCUITS when he shares them.*

- *Buy* Sugar *and* Bliss *magazines to learn tips on how to attract boys.*
- *Use Mum's hair gel for shine.*

I didn't know then what I know now, so whenever I saw beautiful women on the TV their lives always seemed so glamorous and exciting. They were instantly forgiven for any awkwardness, any defects in character they might have possessed. They didn't have to try, it seemed. Neither did the popular girls at school. They didn't have to be nice. How I longed for this privilege, which seemed so far from my grasp. It hadn't yet occurred to me that they, in their way, were invisible, too.

When I became a teenager, something shifted, as though overnight. It happened during the summertime while I was on the way to call for my friend. I was thirteen, on my usual route to her house. Two builders whistled at me in the street. It made my day. Week! Did this mean I was pretty?

That year, older men started to pay attention to me in the street. They would honk their horns or wink or make comments. Was I pretty?

I was fourteen; a man in the next town told me I should be a model and to pose for him.

Was I pretty?

Of course, so desperate for validation was I that I jumped at the opportunity to prove this to myself.

I was fourteen; an older boy convinced me I was pretty enough for him to spend the night with, even though I wasn't sure.

I was fifteen; my boss propositioned me at work. I was pretty.

I was fifteen; I got drunk (which was my favourite thing to do at the time) and kissed my best friend's boyfriend, because he told me he thought I was beautiful and that he had liked me all along. This was a prime example of one of the first times the need to be seen overpowered my moral inclinations. I had already begun to look around desperately, in a need to fill what was missing, which was endless – and to mistake attention for love, which was catastrophic, and had very lasting effects. And so I walked into all of it. All the situations which were always going to happen, so primed and empty of self-esteem that I was. So attention-hungry. All to prove to myself that I was daring, sexy, confident. Which wasn't the truth.

I was sixteen; a married man wanted me to be his girlfriend. I was so, so pretty.

The times when I felt my lowest were the times when I was the most vulnerable. Falling into hell didn't matter so much if I felt myself in proximity to beauty. It seemed to cover up the sheer ugliness of anything else. It meant changing my reality, or at least the face of it – from losing weight to getting more [modelling] jobs, or saying yes when I was being pursued and didn't know how to say no.

The beauty myth – it might be the largest lie we are ever told. The lie about size. The lie about race and weight and gender. I only really became free of my own addictions, disorders and insecurities when that became clear to me. When I realized the extreme beauty of being exactly who you are. I have been a writer since childhood, but I stopped myself for years. It was never that I didn't think I had anything to say, it was because I could not equate it to beauty, to attractiveness, and I took my true creative passion for granted because I could not see the quick fix in it. Where was the importance? Time and time again it was abandoned, in pursuit of another type of validation – things that I thought made me feel better.

These days, beauty lies in something else entirely for me. It lies in honesty and transparency. It lies in the truth of things and the transmutation of lived experience, terrible or not. I think the largest

THE BEAUTY MYTH – IT MIGHT BE THE LARGEST LIE WE ARE EVER TOLD. THE LIE ABOUT SIZE. THE LIE ABOUT RACE AND WEIGHT AND GENDER. I ONLY REALLY BECAME FREE OF MY OWN ADDICTIONS, DISORDERS AND INSECURITIES WHEN THAT BECAME CLEAR TO ME.

disservices we do to ourselves, to our souls, can happen in the name of beauty. Perhaps only because we have misinterpreted it. When I think of beauty now, I think of sharing, of being more than enough, of being of service. I feel most beautiful when I am laughing, when I am making someone laugh, when I am creating, when I am helping anyone create. When I am sharing and speaking to experiences that we all have. There is an inner light that comes through when we step into our true power, not one that is superimposed, stretched into or painted on. It radiates. It's greater than a glow – it's incandescent.

HANNA IBRAHEEM

on the complexity of our relationship with facial hair

Hanna Ibraheem is beauty writer at *Stylist* magazine and stylist.co.uk. After an MA in magazine journalism at City University, she worked at Get the Gloss and *Good Housekeeping* before joining *Stylist* in June 2018. She was shortlisted for Rising Star at the 2018 Johnson & Johnson Beauty Awards and for Most Promising Student (Postgraduate) at the 2014 PPA New Talent Awards. Hanna has written about everything from why people in Japan live longer to the link between posture and mental health.

The first time I became aware of it was when I was nine. During a falling-out with a schoolfriend, she placed her index finger across her upper lip and walked away. At first, I was confused. 'What was that about?' I wondered. As a kid, I'd never really been bothered about my looks and, thankfully, the absence of social media in 1999 meant that my peers and I didn't grow up with any overwhelming pressure on our appearance.

But later that day, after school, I took a closer look in the mirror. Scattered across my upper lip were tiny spikes of black hairs, all different in length. I'd never noticed them before. But once I saw them, I kept thinking about them. As the days went on, I sneaked a quick look at my friends' upper lips. Some of my Asian friends were similar to me, but it didn't really fill me with any comfort; I just wanted to get rid of the hairs. Despite knowing it was silly, it bothered me at the time. Worried that people would stare at my upper lip when I spoke to them, I developed a habit of speaking fast so that conversations would be over quickly. I don't know why, but I was embarrassed about my facial hair, as if I had done something wrong by involuntarily growing it. Looking back now, it seems ridiculous. But at the time, it made me feel self-conscious and awkward.

Even though I was desperate to get rid of the stray hairs, my mum wasn't keen on me dipping into a waxing pot. It was a topic she never brought up and, when I did, she assured me that it wasn't something I should worry about. In fact, she told me I didn't even need to do it. And I'm grateful, because her attitude towards facial hair was a refreshing change from the remarks you'd hear at school. Boys made cruel comments, and girls were even worse. I once saw two girls arguing in the playground, before one screamed at the other, 'Go shave your moustache!' It was shocking. As the years went by, I noticed that more of my friends – especially the Asian ones – began to bring up hair removal in conversation. It even filtered into our lessons. I'll never forget one PSHE class in which we were told about the ways that women develop body hair during puberty and then the teacher went on to list ways that we can remove it. At the time, I thought it was normal. Looking back, it makes me sad.

Friends who had developed facial hair began coming into school with their upper lips, cheeks and chins stripped of even the smallest follicles. And then I had my turn. In Year Seven, my mum let me wax my upper lip at home. But my reaction surprised me. Despite having wanted this to happen for ages, I was equal parts excited and scared. Would it hurt? Would it break the skin? Would it make me bleed? Would my skin react? I hadn't considered the unavoidably painful side of the process and, because of that, the fear was an unexpected emotion. I didn't want to risk removal creams, as my older sister had a bad reaction once and it had put me off ever using them.

As the wax warmed up, I couldn't take my eyes off the pot. The golden substance inside the Veet jar would rid me of the hair that caused me to sit with a hand over my mouth constantly during lessons.

My mum dipped a wooden stick into the jar and stirred the wax. Once she saw that the colour of the wood didn't change, which would indicate that the wax was too hot, she lifted the stick out of the jar. Wax dripped down in long lines like pieces of shimmering thread – it looked pretty beautiful. How painful could this really be?

Starting with the left side, I moved my tongue up over my teeth and gums, making the skin stretched and ready. My mum swept the stick over a section of my upper lip, and I braced myself. The warm wax felt deceptively comforting against my skin and, for a second, I let my guard down. She patted a piece of material over the wax and, counting down from three, pulled the material off in one swift motion . . . and HOLY HELL. I felt a sudden rush of pain in that one small area, but then it disappeared as quickly as it had happened. I had to reassure my mum that I wanted to carry on. I could tell she hated the fact that I was putting myself through pain to get rid of a few facial hairs, but it felt important to me.

Once we were finished, I couldn't help but look at the slightly gross strips of material. The dark hairs, along with their large, thick roots, had been ripped out and were now stuck in the wax. I ran my finger over the prickly ends, feeling relieved. But it wasn't until I was older that I truly realized to what extent these tiny hairs were the root – no pun intended – of lots of confidence issues. Those pieces of hair have the ability to conjure up so much hostility, anxiety and embarrassment.

But why? Why is something that is so common considered so abnormal? And where does this view come from?

The ideal of women being hairless dates far back. It's believed that women in ancient Egypt used to remove all their body and facial hair; and, at the time of the Roman Empire, having no body hair was associated with class and wealth.

As we grow up, there's an unspoken expectation that women 'should' remove their facial hair when it grows. Society tells us that it's only okay for men to grow facial hair and young boys' first strands of a moustache are celebrated as they become men. On the flip side, women are shamed for letting their facial hair grow, ridiculed for their looks and berated until they do something about it.

Yet, while I know it hurts, costs money and panders to the patriarchy, I can't stop myself from heading to a beauty salon when I notice the hairs have grown long. Years of social conditioning, particularly at school, made it difficult to shift the idea that hair removal was something I should do. In fact, I even switched from waxing to threading once I noticed that wax could break the hair and that threading gave me a 'cleaner' finish.

And that mentality didn't end in the playground. Times where I may have missed a threading appointment didn't go unnoticed by some now ex-friends who thought it was okay to point out hairs. So what if I decided to stay inside during a rainy weekend rather than dragging myself to my nearest threading shop? To some, it appeared to be a crime that I let any facial hair creep even a sixtieth of a centimetre longer. It didn't affect their lives, and I didn't understand why I had to go out of my way to make them feel more comfortable about my face.

And let's be honest: nobody really enjoys visiting those threading shops. While the £2 price tag is great, the noise is overbearing and the chaos unnerving. Women sit on chairs against a wall waiting for one of the beauty therapists to call them forward to take a seat in the chair. When you do, they pull the chair back, leaving you lying almost flat, and get to work. The therapist taps your face, indicating where she wants you to stretch your skin so that she can get to the hair easily. Moving her index fingers and thumbs towards each other, alternating between left and right, the thread dances around her hands while her head bobs up and down. In the haste of it all, I've left the shop with a cut on more than one occasion. When it's over, she dusts the hair off your face, applies a cooling aloe vera gel and sets you off on your way. She then calls the next woman forward, and the line of women waiting is even longer than when you arrived. It's a never-ending line, and a profitable one. While you may drop the odd £2 or £3 here or there, it all adds up. I sat down and worked out that by visiting a threading shop even just once a month, a woman can be set back £1,040 over ten years. In a bid to avoid coughing up the cash – and so that I can avoid the shops when I don't have the energy – I've taught myself how to thread. It's a tough skill to learn. If you don't use the right thread, it can feel rough against your skin and the thread cuts into your fingers.

But the one thing that really strikes me about threading places are the hands working the thread. Most of the employees are Indian, whether you're visiting a fancy salon in Liberty or a local shop. Threading is an age-old method that originated in Iran, India and Central Asia and has been practised for centuries. Over time, the technique trickled into the Western world, but the skill has stuck with Asian women, who continue to dominate the industry.

To me, race and culture play a big part in our personal relationship with facial hair. For most white friends growing up, it was never even a blip

WHILE YOU MAY DROP THE ODD £2 OR £3 HERE OR THERE, IT ALL ADDS UP. I SAT DOWN AND WORKED OUT THAT BY VISITING A THREADING SHOP EVEN JUST ONCE A MONTH, A WOMAN CAN BE SET BACK £1,040 OVER TEN YEARS.

on their radar. Those who did grow facial hair usually had blonde facial hair, meaning it could hardly be seen unless you had a magnifying glass to hand. Meanwhile, my Asian friends and I dealt with darker, thicker and coarser hair. We also, somewhere along the line, all came to the conclusion that Asian women are generally hairier. And it's not just in their heads. Sadly, Asian culture is riddled with complexities around what is 'beautiful'. It's a culture that widely believes fair skin is lovely – so much so that fair-skin creams are commonly found in bathroom cabinets in India. This warped view lends itself to ideas that women should not be hairy – whether it's their upper lip, their arms, their legs, their underarms – just anywhere. And it meant that many of my Asian friends and I were familiar with products like Jolen at a young age.

Popular culture has also played a big part in the way female body hair is viewed in society.

From the trope of the bearded lady, who is considered a 'circus freak' by some, to the outrage around Julia Roberts daring to attend the *Notting Hill* premiere with hairy armpits, we're constantly being fed the message that, unless it's on your head or your eyebrows, hair on women is strange, unusual and abnormal. Even just last year, MAC Cosmetics posted an image on its Instagram page of a lip pencil being applied to a woman's lip. Upon closer inspection, you can see that the woman has a hairy upper lip. Sadly, the picture was inundated with negative comments, highlighting society's distorted perception of beauty.

For me, it's still one of the most refreshing things I've seen on a social network that is flooded with unrealistic beauty ideals. And thankfully, a handful of women left comments on the picture saying they wish they had seen something like that from a big brand when they were growing up. As a kid, having facial hair affected how I behaved.

I retreated into myself to avoid attention. So any movement towards normalizing facial hair on women is a big thing in my book.

Despite these steps, there's still a long way to go. Friends have told me that they apply heavier make-up to their eyes to distract from a hairy upper lip. Others refuse to tie up their hair for fear of revealing their sideburns. It's sad, but then again, while I no longer obsess over it like I did as a teenager, I do still schedule in time for hair removal. Why? More out of habit than anything else. But that's my choice. And when I don't do it, that's my choice, too.

Despite knowing that I pay money and go through pain to fix a problem that isn't really a problem, it's something I can't shake. Feeling self-conscious about it as a young girl has left a mark on me. If I feel like the hairs have grown long and I'm speaking to somebody, I get a horrible yet familiar feeling in my stomach that it's all they're looking at. If they are, I know it's their problem and not mine, but years of social conditioning have made it difficult to go thread-free. However, I have hope that this won't be the case in the future. The mentality towards body hair is shifting, and women are finally embracing what feels best for them. I just hope that, one day, I can learn to do the same.

ROSE McGOWAN

on rejecting the 'princess' hair of Hollywood

As a writer, director, entrepreneur and feminist whistle-blower, Rose McGowan has focused a spotlight on injustice and inequality in the entertainment industry and beyond. As an activist, she became a leading voice in the fight against sexual assault. Rose gained recognition as an actress with lead roles in films such as *The Doom Generation, Scream, Jawbreaker* and *Planet Terror.* She starred in the hit series *Charmed,* one of the longest-running female-led shows in TV history. Her directorial debut, *Dawn,* was nominated for the Grand Jury Prize at the Sundance Film Festival.

I was a born dissenter. My father says I came out of the womb with a fist up, and I suppose it's true. Around twelve, thirteen years old, I started wearing mostly all-black clothes and listening to very moody music. I dyed my hair black and made spider webs on my face with black kohl – needless to say, I did not fit in. I lived in a state called Oregon at this point, and everyone who was very popular at my small-town school had bleached blonde hair, with the puffy bangs at the front. I wanted to cut myself off from what these people looked like or expected me to be. The homogenized idea of how a girl should look annoyed me. These people didn't like me, and I didn't want to look like them.

I went to the pharmacy, bought a box of black hair dye, and that was it. I selected Clairol's blue/black shade; I didn't exactly want to look natural. My hair was kind of a reddish

brown but I loved that really black hair look.
To the other kids, I was this weird girl, a foreigner in a foreign land. I was called ugly every day, to the point where I'd just respond, 'Yeah, yeah, yeah, tell me something I don't know.' Things were thrown at me at school. But that only made me want to stay different. I didn't look ugly to myself, so it was quite confusing. At some point, I moved to the state of Colorado, where, for whatever reason, I suddenly found myself being popular. The kids at school seemed to love my short black hair. It taught me a really good lesson; it made it so I didn't believe the people who said I was awful and I didn't believe the people who put me on a pedestal. I decided to keep my own counsel. I also used so much hairspray I broke my long fringe off accidentally.

After this, I moved to another place: Seattle, in the state of Washington. I met some cool hairdressers from London and got cool hair extensions that were a mix of blue and black chin-length corkscrew curls. They were kind of Betty Boop-esque, and I loved them. Sometimes, when they got loose, I'd pull them out on the bus to scare people just for fun. I loved looking however I wanted to look. It was great, until it changed.

Enter Hollywood. Most of my life, until I came to be in the film industry, I had quite short hair. When I started auditioning for parts, a female agent told me – and this is a direct quote: 'You have to have long hair so that the men in Hollywood will want to fuck you. If they don't want to fuck you, they won't hire you.' I was so young when I got this information. And it stuck like glue. Having previously been homeless, I definitely wanted to get hired, so I thought, *Okay, I guess I have to have long hair.* Being homeless is not fun. I grew out my hair, and it's so interesting to me, how it made me feel. It was like the tale of Samson

‘YOU HAVE TO HAVE LONG HAIR SO THAT THE MEN IN HOLLYWOOD WILL WANT TO FUCK YOU. IF THEY DON’T WANT TO FUCK YOU, THEY WON’T HIRE YOU.’

but in reverse: my power faded away, the more hair I had. Even now, when I put on a long wig, I feel the same. There is something about what it does to my face, and even my body changes radically. Instantly, I feel like a false version of myself. This was strange to me because, for so long, I had captained my own ship, so to speak.

Later, when I was on the TV show *Charmed*, we three lead actresses had dark hair because we were sisters on the show. One day I dyed my hair red because I didn't like blending in. I didn't like that we had to get the approval of the studio head if we wanted to change our hair. My contrarian nature flared up. I just did it. The studio panicked, wondering, 'What are we going to say? How will the audience know who she is?' Luckily for me, because it was a show about magic, I came up with an easy and fantastical explanation: 'A potion blew up in my face and turned my hair red and I liked it.' Lo and behold, those were the first lines of the next season.

Most women don't get told directly, 'You must have long hair,' but society tells them that in a million small, insidious ways. It's so deeply ingrained. Hollywood gives women a distorted mirror to look into, and I was one of those mirrors. I was the one sent to make you not feel good about yourself when you're sitting next to your boyfriend in the theatre watching me.

Hollywood affects your mind and beauty standards in ways that you may not even be aware of. The same thing happens in the media industry, and in advertising, but Hollywood is particularly devious. It's on purpose, it's calculated and they do a really good job.

For me, my hair was always tied into job security, and I resented it. I always had my hair blow-dried, because then I wouldn't have to deal with it for at least a few days. I thought that getting my hair blown out was the closest I could get to living in a more effortless fashion, like a man gets to. It's a strange thing when you are packaged and marketed and sold as a sex symbol. It ostracizes you from most people. Many women scorned me, and many men thought they could touch me. Living in that blown-out look, I think I became immune to my external self; I could no longer see what I looked like. Looking back, I'm actually embarrassed it got so out of hand. The hairdressers just kept making it bigger, and I just failed to notice. This went on for years. That is what I finally looked at. I looked at my hair. I looked at myself. I looked at why I had the big hair on my head. And I decided I was done.

One day, I had had enough. Enough with being a brunette Barbie doll. When I shaved my head there was no moment of freaking out, no moment of, 'Oh my god, I need my hair back.' It was simply a moment of total freedom. It was beautiful to me, and it felt great. Now, maybe I am not as attractive to the traditionally minded. But, so what? Why would I want to fit in that paradigm anyway? I tried it, and it didn't work for me. Maybe it doesn't work for you either.

Right now, you might be thinking, *I like my long hair*, which is great. Everyone should look however they want to look. Just check in with yourself and decide why. Do you look like what you want to look like, or are you adhering to society's rules of what you're supposed to be? If you go down the rabbit hole and look at the end result, it's often that the pressure of having to be societally attractive is involved. Or maybe you simply like longer hair – just do a check-in with yourself and see. There are many women who have breast implants, artificial nails, hair extensions, fake tan . . . all of these things are fine as long as it's truly for themselves. Who is dictating worth? Is it you, or someone else? Is it you or social pressure?

Now, I wear make-up and I dye my hair, but in a playful way. If I feel like I'm in a red-lipstick mood, then that is very different from a pink-lipstick mood, and how great that we can have fun with that. I look at everything as an accessory to myself, and for myself, not for anybody else.

What I noticed when I shaved my head was that men, specifically, could hear me for the first time. I never felt heard before. They actually acknowledged the words coming out of my mouth, and I realized it was because I was glitching the system by looking different. I wondered if I could help women be heard without them having to shave their heads.

I meet so many women who tell me, 'I hide behind my hair.' And I ask, 'What are you hiding from? Are you hiding from your own power?' For me personally, I came into my power once I stripped all my hair away. For women to choose to have a shaved head is still unsettling to many people. I get loads of messages on social media telling me to grow my hair back. But grow it back into what? Docility? To avoid others' discomfort? No, thank you very much for chiming in, but no thank you.

Change is happening. I believe people will look back on this time and see a seismic shift. Women are being heard at last, men are speaking up about being hurt. People come up to me in the street to tell me they are being listened to in meetings, for the first time ever. Just recently, I overheard a man say something rude about a woman and the man with him said, 'That is not actually a good way to speak about humans.'

It heartened me. Women also need to look at their part in this system, because they can be complicit. And if you are complicit in a structure that doesn't benefit you, guess what you are in? A cult of thought. Of course, most people don't think they live in a cult, they may think they live in a free society. But look at the people who hold the power, and look at the messages that keep people down, especially women. Simply by nodding your head and going along with it, what part are you playing in this situation?

I do notice boots-on-the-ground change. To me, the movement is about consciousness-raising and changing people's perceptions of what they are and who they can be. Because we can be all of it. We're not just limited to what we are allowed to be because we were born into something over which we had no control. Women are realizing that they have power. Many men are going through a process of assessing where that grey area is and thinking twice, which is a good thing. If we can achieve even one less hurt human on this planet, that's a great thing. And if we can achieve it with the hair that we want, then that's even better.

CHIDERA EGGERUE

on owning your womanhood

Chidera Eggerue is an award-winning blogger and author with a fast-growing audience of over 100,000 across her social media accounts. She grew up in Peckham, London, and her blog, The Slumflower, was inspired by the idea of a rose growing from concrete. After an appearance on ITV's *This Morning*, Chidera sparked a conversation about body positivity, which went viral with the internet-breaking movement she created called #SaggyBoobsMatter. She's on a trailblazing path to encourage us all to be more comfortable with our perceived 'flaws'.

I have always felt too small to be seen, yet, at the same time, too large to be loved. I've always felt like I was asking for too much from the world when, actually, I was just asking the wrong people. I wish someone had told me that it's not my responsibility to be desired. Why didn't they tell me that my purpose extends far beyond being liked? Or that it isn't the end of the world if I don't meet someone else's standards? I guess I had to discover that truth by myself. And I've learned a lot along the way.

It's impossible to spend time on this planet without its corrosive ideologies rubbing off on you. For some of us, it takes a lifetime to purge ourselves of them. Sadly, many of us may not ever know what a life that isn't wrapped around self-hate looks like. There is a higher version of every single one of us waiting to be accessed, but the rigid systems of society, carefully crafted to keep us reliant on external power, block us and bind us to toxic ideals instead.

For anybody who has boobs, you'll probably be familiar with the feeling of disliking your breasts. My boobs used to be my biggest insecurity, because they're saggy. That's right: I have saggy boobs. They hang low and swing like pendulums.

My body matured faster than my mind did and, ever since I was a young teenager, boys made me hate my body. I used to get called names like 'slipper boobs', which drove me to the unnecessary measure of trying to buy push-up bras at such an early stage in my life. I remember going to Marks and Spencer to get my bra fittings done and, each time I tried on the correct bra according to my accurate chest measurement, I'd stare at the mirror looking at my deflated breasts with disappointment. I would unfavourably compare myself to the model on the packaging, smiling, with her perfectly perky cleavage. But it wasn't the model's fault that I developed an inferiority complex. Society had failed me.

I remember, when I was fifteen, I told my mother that I would like to get a boob job as soon as I turned eighteen. I hated my body because the world had told me to hate it. Thankfully, by the time I was eighteen, I had woken up to the idea that it wasn't actually me who needed to change, it was society. I was struck by how ridiculous it was that, even as a teenager, I was already being made to feel that my breasts were not round and pert enough. I didn't go through with the procedure (I'm not sure how I thought I was going to afford it, anyway); instead, I decided to begin a mental investigation: who taught me to hate myself, and why do I believe them? The answer was and still is the patriarchy – the system that the world operates on, which allows men to have disproportionate access to power, money and resources. At the age of nineteen I decided to stop wearing a bra and to talk about my decision honestly on social media. This was met with a lot of cyberbullying and cynicism. I powered through and continued to put myself first until my early twenties, when I realized that there is a much larger conversation to be had about women's bodies and the idea that we exist to be consumed by men.

That's when #SaggyBoobsMatter – the hashtag and campaign that I launched for women of all shapes and sizes to celebrate our bodies – was born. That campaign, and my blog, The Slumflower, opened up an important debate around women's bodies but also about feeling more confident, especially as a black girl.

I'm twenty-four years old now, and I don't like wearing bras. They make me feel restricted and squashed. So I have opted out of doing so. Small-breasted women are 'allowed' to go braless because there is less to judge or make people feel uncomfortable. There is a stigma attached to not wearing a bra if you have big boobs, but that stigma is yet another patriarchal attempt to control women. I'm not here for it. Fight me.

One of the biggest lessons, and one I'm still learning, is the importance of owning your beauty. My mind was thoroughly blown when I came to the realization that I decide how beautiful I am, because to be desired by myself is the deepest adoration of all. Being beautiful is subjective. Some would describe beauty as external and others would describe it as internal but, for me, my beauty comes from my unwavering ability to own myself – to stand tall and firm in my truth and to reject any ideas that push me closer to thinking less of myself. I have been wonderfully made, with beautiful veins knitted together to form this fragile yet resilient being who is more than just an assortment of facial features. I am a walking capsule of stories, a reminder that my ancestors made a set of unknowingly divine decisions that led to my arrival. I must honour that. I must honour me.

From the moment a woman is born she is taught that she's not good enough and that she must dedicate the rest of her life to meeting everybody's standards but her own. She is taught that she's less of a person and more of an object that is to be consumed, judged and discarded once her assigned value expires. I guess a world run by cisgendered heterosexual men (also known as the patriarchy) can operate only if women hate themselves. In order to control someone, the trick is to convince them not only that they aren't good enough but that you have power over them so that they will be distracted by the endless pursuit of the power-holder's validation.

When this clicked in my mind it was like a light bulb had exploded above my head. I said to myself, 'You mean to tell me that hating myself is a choice and that I can actually embrace the body I've been born into without any repercussions? Sign me up!' So I signed up to the self-love club, and my membership is for life. The best thing about this club is that every single person is welcome – including you. There are no entry requirements: all you have to do is arrive as you are, but you must be willing to be honest with yourself.

Regret is the greatest teacher. You don't learn anything when you're right, but you learn a hell of a lot when you're wrong – in fact, learning can come only from mistakes. But what about the ones we cling on to? For a long time, I regretted not wearing bras to sleep, as I was advised to by misguided women. Thank goodness I chose my delightful sag over the discomfort! As a recovering overthinker, I've had to learn the hard way that pondering on what could have been will only delay what's coming. I regret spending so much time worrying about my saggy boobs and planning how I would change them, intending to have surgery. I could instead have been living my life in a positive way and embracing the confidence that came with the realization that women do not exist in order to be consumed by men.

Revenge is never something that is worth seeking. Because when people fail to value and protect the opportunity of having us in their lives, they must live with the regret that comes with their own (eventual) growth. The funny thing about others taking advantage of delicate people like us is that it's all fun and games until their value system changes and they've run out of opportunities to rectify the past. People regret treating you badly when you start to treat yourself better. The most hilarious thing about people who have wronged us reappearing in our lives is that, by the time they return, we've already done our healing and have moved past waiting for them. If you've managed to find clarity at the end of all the chaos, you have won.

There is something my incredible mother taught me that will follow me for ever throughout my life, and it is that silence can never be misquoted. Some things just don't deserve a reaction, especially situations that will only drag you back to that person you've made the commitment to grow out of being. I had to learn to ignore negative comments and not rise to the bait of online trolls. I know that most negativity comes from people who are trapped in their own insecurities, or people who are intimidated by my confidence in not wearing a bra when the under-representation of saggy boobs in the media has led them to believe there is only one way to be beautiful.

Choosing silence is a form of stillness that can be mastered only through understanding that, sometimes, you don't need to win; you just need to do the right thing. It's so tempting to give in, but why try to live up to the expectations of people who don't even meet their own expectations of themselves? A lot of the time, the 'right' thing looks like losing because it's often achieved through a refusal to rise to the bait of a person who is struggling with their own cocktail of problems. When people are unhappy, they live through destruction because it gives them a sense of power that they lack elsewhere in their life. Stillness maintains inner peace. Having said that, if something has gone on for too long and you know you deserve to be in a better environment, then move on. You will thank yourself for doing so.

Maintaining stillness and protecting your new-found peace can only work if you implement boundaries. This word scares a lot of people – especially women. It's almost as if we're not allowed to say no, but if feminism could be compressed into one word, it would be 'No'. Boundaries are about distance and rules – something a lot of people struggle to grasp, let alone respect. Boundaries only threaten people who already feel like they have an entitlement to us. The less you tolerate, the more people will respect you. The more you tolerate, the less people will respect you, because there is no line drawn that they can't cross. This isn't to say that it's your fault if people choose to treat you badly. Rather, it's a reminder that you are your own keeper and you are your own responsibility. If you're too scared to protect yourself, think about protecting the four-year-old version of you. That child still lives in you and is the source of all the emotions you feel, including the ones you run from. Listen to her and look out for her. She needs you.

Being sensitive is wonderful because it allows you to connect with others, but where it can become problematic is the moment you allow these feelings to alter the way you view not only yourself but the world around you for the worse. When I was a teenager, insecurity about my body meant that I always had a voice in my head telling me that I wasn't good enough. It's so easy to let other people's voices creep into our minds, but what

IF FEMINISM COULD BE COMPRESSED INTO ONE WORD, IT WOULD BE 'NO'. BOUNDARIES ARE ABOUT DISTANCE AND RULES – SOMETHING A LOT OF PEOPLE STRUGGLE TO GRASP, LET ALONE RESPECT. BOUNDARIES ONLY THREATEN PEOPLE WHO ALREADY FEEL LIKE THEY HAVE AN ENTITLEMENT TO US. THE LESS YOU TOLERATE, THE MORE PEOPLE WILL RESPECT YOU.

works for me is reminding myself: nothing anybody ever ultimately does is because of me. Those voices used to tell me I was ugly and unworthy – thank goodness I never listened. People are mean because someone else has been mean to them. Their interaction with me is a choice that has been influenced by the way they interact with their own self.

Every day, my aim is to move closer and closer to the highest possible version of myself. I am learning to be gentle, patient but, most importantly, prepared for the woman I am about to become. She is going to need all the strength I have in me to move through a world that will continue to fight to shrink me. But I believe in her. And I believe in you. We've got this.